ORDERLY TANGLES

ORDERLY TANGLES

CLOVERLEAFS, GORDIAN KNOTS, AND REGULAR POLYLINKS

ALAN HOLDEN

with photographs by
DOUG KENDALL

Columbia University Press
New York   1983

Library of Congress Cataloging in Publishing Data

Holden, Alan.
  Orderly tangles

  Bibliography: p.
  Includes index.
  1. Polyhedra. 2. Knots and splices. 3. String
figures. I. Title.
QA491.H626 1983   516.2′3   83-1780
ISBN 0-231-05544-7

Columbia University Press
New York   Guildford, Surrey

Clothbound editions of Columbia University Press books are Smyth-
sewn and printed on permanent and durable acid-free paper.

FOR JAYNET

Beloved Helpmate

CONTENTS

Preface   ix
1. Highway Interchanges   1
2. Nets, Knots, and Noknots   9
3. Gordian Knots   21
4. Cat's Cradle   35
5. Chained Polylinks   45
6. Regular Polylinks, 1   53
7. Regular Polylinks, 2   63
8. Regular Polylinks, 3   71
   Appendix   85
   Bibliography   93
   Index   95

PREFACE

The words "orderly" and "tangle" seem at first contradictory. But traversing the tangle of roads at an airport becomes orderly with the assistance of directive signs. And "the tangled web we weave when first we practice to deceive" is orderly in intent.

In fact intent is the bridge between "orderly" and "tangle." The tangle of roads acquires order in light of need to get us all where we want to go with a minimum of mutual interference and with maximum dispatch. The "cloverleaves" at the intersections of major highways have a similar intent, as the first chapter of this book illustrates.

That chapter introduces another idea important in practical affairs, the idea of redundancy. Applied to highway interchanges, redundancy may sound at first like a mistake, a needless expense to the taxpayer, but motorists would complain loudly if they were forced to traverse nonredundant highway interchanges.

Probably man's oldest purposeful tangles are those of weaving and its ancestor, basketry (chapter 2). Knotting, too, must have appeared in prehistoric times. The mathematician's knots (chapter 3), with no open ends, did not receive notice until the nineteenth century. But the celebrated knot, treasured by King Gordius of Phrygia and hacked open by Alexander the Great, has this character.

Gordian knots call attention to the distinction between knots and rings. Often a ring can appear to be knotted, as anybody who plays "Cat's Cradle" with a loop of string knows (chapter 4), and further complications in the distinction between rings and knots appear when several rings are linked together into chains (chapter 5). It is not a purpose of this book to pursue such distinctions.

Most importantly the book puts on record the idea of "Regular Polylinks," and pursues some of its consequences in chapters 6–8. This material grows out of work of the classical geometers on polyhedra (see *Shapes, Space and Symmetry*, New York: Columbia University Press, 1971). But it has received earlier attention only by this writer (*Structural Topology* #4, Montreal, Quebec, 1980).

Alan Holden

ORDERLY TANGLES

1

HIGHWAY INTERCHANGES

Automobile drivers often refer to the tangle of roads at a major intersection as spaghetti. It forms a dish that becomes more digestible, if not more palatable, after thinking about the several problems that highway engineers face where major highways cross one another.

They know that motorists who intend to stay with one of the major highways through the interchange expect to be able to drive straight ahead without slowing down. Motorists who wish to turn from one highway to another expect to be able to do so without interrupting or being interrupted by other traffic flows. And finally, at best, motorists who wish to make a U-turn expect to be able to turn around within the confines of the interchange.

Here, then, is the outline of the engineers' general problem: to provide connecting roads which will fulfill these expectations. In almost any particular case they must solve other more specific problems also. For example, difficulties often arise from the amount and shape of the space that can be allotted to the interchange.

The first step in analyzing the general problem is to look upon each major highway as consisting of two separate roads running parallel and restricted to conducting traffic in opposite directions. Each sub-highway must be given access to every other by a one-way passage, which may consist of a single road or a succession of fragments of roads.

The most familiar example of such an interchange is the cloverleaf, often found where two major highways cross. The name may seem an odd one considering that characteristically the traffic cloverleaf has four lobes, whereas the botanical cloverleaf has three. But Nature generously provides an occasional four-leaf clover to bring good luck. And three-lobed highway interchanges do occur.

One highway passes over the other, so that no sub-highway intersects another. All turns off sub-highways are right turns and all entrances are right entrances. Thus, except at times of congestion, all traffic, even that moving from one sub-highway to another, proceeds without interruption.

One might expect that the task of connecting each of the four sub-highways to each of the three others would require twelve one-way connecting roads. But in fact only eight connections are usually provided. The four omitted connections are those that would directly permit U-turns.

A motorist can make a U-turn at the interchange, and it is instructive to see how. First the driver turns right onto a lobe of the cloverleaf and reaches the other sub-highway. After a short jog along it, he turns right again, along a second lobe of the cloverleaf, and so reaches the sub-highway running opposite to the sub-highway that he originally left. In effect the driver divides the half-turn that he wishes to execute into two quarter turns.

Notice that by making another quarter turn using a third lobe of the cloverleaf, the motorist could reach the sub-highway that he could have reached directly from his original sub-highway. In other words, if the motorist had missed the turnoff which would have taken him directly to where he wanted to go, he could have gone there by taking three other connecting roads within the same cloverleaf. Thus the cloverleaf provides more connections than are strictly necessary. A mathematician might call it "redundant."

When all four of the unnecessary connections are removed, only the four connections forming the four lobes remain, as fig. 1.1 shows. It is important to notice that, by traversing in succession all four lobes, the motorist would return to the sub-highway from which he started, after briefly visiting the other three sub-

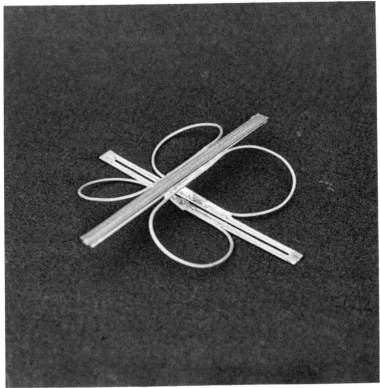

Figure 1.1 The nonredundant interchange between two major highways resembles the familiar cloverleaf shorn of its four right-turn connections. A motorist can still make a right turn by traversing three lobes of the cloverleaf.

highways. Again therefore, you see that the nonredundant cloverleaf, with four lobes, forms a *sufficient* interchange for the four sub-highways. Dividing the two major highways where they cross each other enables the nonredundant cloverleaf to accomplish its task even more smoothly (fig. 1.2).

Clearly a satisfactory *practical* interchange between more than two major highways must entail some redundancy. The best form for it to take can be decided by returning to the nonredundant interchange (figs. 1.3–1.5) and asking what connections to add in order to reduce from five to four the maximum number of connections to be traversed by any car where three highways meet.

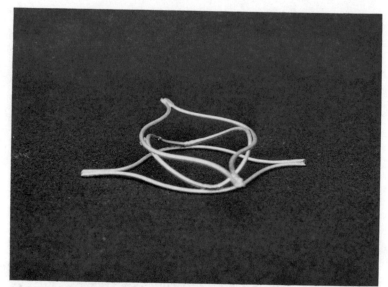

Figure 1.2 The nonredundant interchange between two major highways can have a radically different appearance from the cloverleaf (fig. 1.1) and still accomplish the same tasks. This design suggests the close relation between the interchange and a traffic circle.

Figure 1.3 This nonredundant interchange for three major highways is again reminiscent of the familiar cloverleaf (fig. 1.1). It has four lobes and two additional connections.

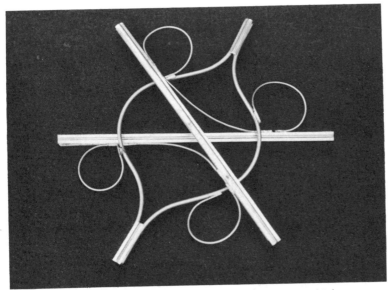

Figure 1.4 A nonredundant interchange for three major highways acquires a new flexibility in its design by dividing one of the major highways.

Figure 1.5 This nonredundant interchange between three major highways is still a third example of how the connections can be placed.

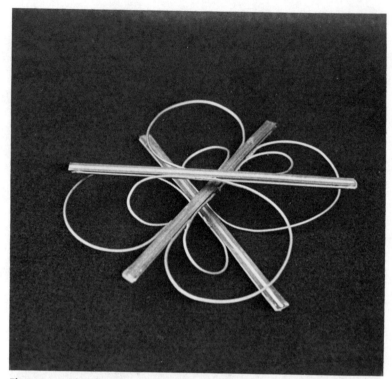

Figure 1.6 This redundant interchange between three major highways provides no single circuit connecting the six sub-highways. Nevertheless a motorist can reach any desired sub-highway by using a maximum of four connections.

After a few trials you find that there is a way of adding only two more connections that will enable all cars to reach their desired directions using no more than four connections (fig. 1.6).

Similarly you can find a way of adding three connections that will reduce the maximum of required traversals from five to three. And six added connections can reduce that maximum from five to two (fig. 1.7). Perhaps this last result—a three-highway interchange with twelve connections and a maximum traversal requirement of two—would provide an acceptable compromise for many cases in real life. The reduction in the number of connec-

Figure 1.7 Twelve connections between the sub-highways reduce to two the maximum number of connections that any motorist must traverse.

tions from thirty (for complete redundancy) to twelve would effect a large simplification in engineering.

In tracing possible courses of automobiles through these interchanges or designing alternatives it is important to bear in mind two requirements. On each sub-highway and each connecting road automobiles travel in only one direction. In addition, on any sub-highway all exits from connections must precede all entrances.

The redundancy of the connections at an acceptable highway

interchange is a conspicuous example of the importance of redundancy in almost all aspects of practical life. For example, most
humans leading civilized lives devour much more food than they
require. Humans introduce into their systems of communication
much redundancy to make sure that they will be understood un-
ambiguously.

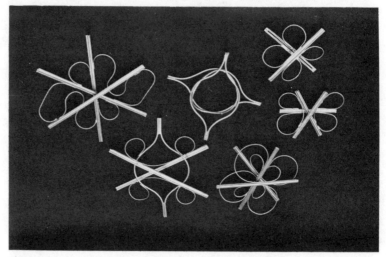

Figure 1.8

2

NETS, KNOTS, AND NOKNOTS

Weaving and its ancestor, basketry, are surely among mankind's earliest efforts to tangle things to a useful end. Ancient baskets were woven on a base of thick reeds laid parallel to one another by passing thinner reeds over and under alternate reeds in that base. When chinks in the resulting web were filled with grass and clay, the basket could even hold water.

In the later woven counterparts of this construction a "warp" of threads is held parallel, and a "weft" thread is passed alternately over and under each. The weft thread is then passed back through the warp in similar fashion, so that its overpasses fall against the underpasses of the preceding traversal. The thread is beaten against its predecessor and the whole operation is repeated to form a dense web of interlacing threads.

Figure 2.1 shows at the left a wooden model of the course of the threads. Weavers call such a web "tabby." By varying the colors and weights of the threads in the warp and the weft, tabby can be made to take an astonishing variety of appearances.

A simple extension of the idea of tabby gives rise to the weave called "twill." Here each weft thread passes alternately over and under two warp threads, producing the web suggested by the wooden model shown at the right side of fig. 2.1. The material so woven may look like a dense array of chevrons.

Even experienced weavers often do not recognize that twill

Figure 2.1 In the simplest weaving configuration, "tabby" (left), two families of parallel threads lie perpendicular to each other. The threads of each family pass alternately under and over those of the other. When each thread passes alternately over and under two (right), the weave is called "twill."

can be analyzed into an interpenetration of two independent tabbies. On this point wooden models such as those pictured in fig. 2.2 can be especially revealing. The twill at the right is a compound of two tabbies such as that shown at the left. If each tabby is rigidly glued within itself, but is not glued to its companions, the tabbies can be moved about as rigid wholes.

In tabby and in twill two families of parallel threads interlace each other perpendicularly, and weaving in the strict sense is limited in that way. But there are other means of interlacing threads and it is instructive to examine some properties of the networks they can produce. Figure 2.3, for example, shows a wooden model of the course taken by three instead of two families of parallel threads intersecting at an angle of 60 instead of

Figure 2.2 The twill weave (right) is composed of two tabbies.

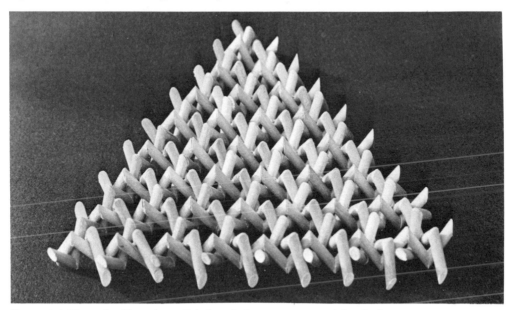

Figure 2.3 Three families of parallel threads form a net containing both hexagonal and triangular meshes.

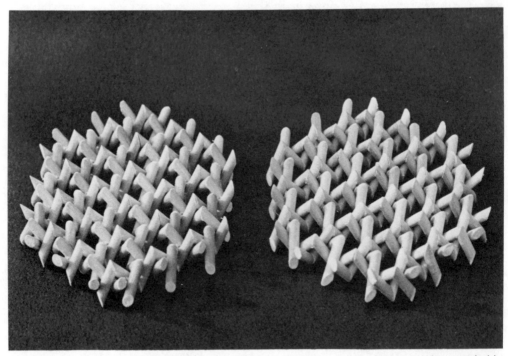

Figure 2.4 The network of fig. 2.3 can be made in two distinguishable forms which are mirror images of each other.

90 degrees. In place of the square meshes in the tabby net, there are here hexagonal and triangular meshes.

An examination of this weave in close detail reveals a further and more fundamental complication. Following the courses of the three threads that participate in forming any one triangular mesh, in a clockwise sense around the mesh, one finds the sequence of crossings to be under-over, under-over, under-over. But it is quite possible to make a net in which this sequence is reversed, as fig. 2.4 shows.

Notice that these two nets, though so much alike, are in fact quite distinguishable from each other. Neither can be turned into the other by "turning it" in any way, even by turning it over, front to back. They are like a right and a left hand, similar but not identical. Either can be made to look like the other only by reflecting it in a mirror. They are described by mathematicians as

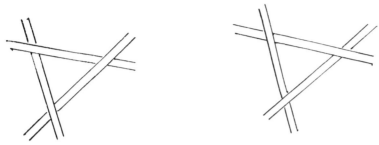

Figure 2.5 The triangular meshes in the networks of three families of parallel threads may take either a right-handed or a left-handed form.

"enantiomorphs" of each other, a word derived from a Greek root meaning "opposite" (fig. 2.5).

What is it that enables these triangular weaves to be made in enantiomorphous forms while the square tabby weave cannot? The ability is best understood by examining some general principles of geometric symmetry. Indeed those principles will turn out to be useful in other contexts in this book as well.

The ideas of geometric symmetry are made precise by specifying the operations that can be performed on an object that leave it looking as it did before their performance. For example, in the cube diagrammed in fig. 2.6 the line A, passing through the centers of a pair of opposite faces, marks a direction around which the cube can be turned through 90 degrees and still look as it did before. This direction is called an "axis of fourfold rotational

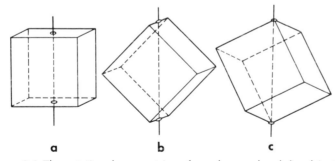

a   b   c

Figure 2.6 The rotational symmetries of a cube can be defined in terms of three axes of fourfold rotation of the sort A, six twofold axes B, and four threefold axes C.

symmetry," because the cube can be turned about it into four successive positions during one revolution, in which its appearance is the same. Each of the three pairs of faces of the cube affords the same possibility, and the cube is therefore said to have three axes of fourfold rotational symmetry.

In a similar way the line B, passing through the midpoints of a pair of opposite edges, marks an axis of twofold rotational symmetry. The twelve edges of the cube come in six opposite pairs, furnishing six axes of this sort. Finally, line C, passing through a pair of opposite corners, marks an axis of threefold rotational symmetry, and the four pairs of corners provide four such axes.

The cube also admits other symmetry operations of a radically different kind from the rotational. These operations are suggested in the diagrams of fig. 2.7. A plane such as that at A, cutting the cube in half, is a "plane of reflection symmetry." If all parts of the cube were moved perpendicular to that plane, through it, and out an equal distance on the other side, the cube would look the same as before.

Evidently the cube has three mutually perpendicular planes of reflection symmetry of the sort described at A. Moreover there are six planes of reflection symmetry of the sort shown at B, pass-

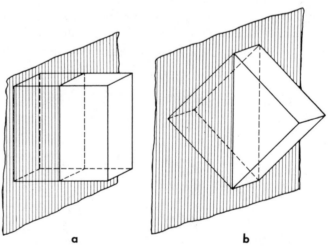

a   b

Figure 2.7 A cube is symmetric to reflection through three planes of the sort A and six planes of the sort B.

ing through pairs of opposite edges. In sum, the symmetry of the cube can be defined as consisting of six axes of twofold, four axes of threefold, and three axes of fourfold rotational symmetry, and nine reflection planes. A single reflection plane, passing vertically through the body, describes the external symmetry of a human being.

It is important to notice the deep-seated difference between an axis of twofold rotational symmetry and a plane of reflection symmetry. Superficially it might seem that rotation through 180 degrees and reflection through a plane accomplish the same result, turning a thing back to front. But everyone knows that a left-footed shoe cannot be turned into a right-footed shoe, even though the two shoes are mirror images of each other. Rotation and reflection are far from equivalent. An object can be rotated by picking it up and turning it, but it cannot be reflected without taking each of its parts and moving them through each other, or by some optical trick such as using a mirror. The first operation is "performable" and the second "nonperformable."

Herein lies the difference between things that can appear in distinguishable enantiomorphous forms and things that cannot. Since two enantiomorphous objects are mirror images of each other, reflection converts either into the other. If an object has a plane of reflection symmetry, another reflection will only interchange the parts on either side of the symmetry plane, and thus leave its appearance unchanged. In other words, only an object lacking a plane of symmetry can appear in enantiomorphous forms. An object with a plane of symmetry already contains its own mirror image, so to speak.

There is another important nonperformable symmetry operation, inversion through a center, which can be regarded as a combination of two of the preceding operations. After a center has been identified, inversion through it moves each point of an object along a line through that center and out to an equal distance on the other side of the center. If the object looks the same after this operation, it is said to have a "center of inversion symmetry."

Recognize that the appearance of a center of symmetry in an object does not imply that the object has both a plane and an

Figure 2.8 The figure-eight knot is a familiar example of an object whose symmetry is completely described by a center of inversion.

axis of symmetry. The composite can appear quite independently. For example, the model of a figure-eight knot pictured in fig. 2.8 has a center of inversion symmetry at the middle of the horizontal crosspiece. But it has no plane of reflection symmetry and no axis of rotational symmetry: the center is its only symmetry element.

The models appearing in fig. 2.9 offer an interesting contrast between man's two simplest knots: each has just one symmetry element. In the figure-eight knot at the left, that element is the center of symmetry just described. In the overhand knot at the right, that element is an axis of twofold rotational symmetry. Since the overhand knot lacks nonperformable symmetry elements, it, unlike the figure-eight knot, can be tied in two distinguishable enantiomorphous ways.

More complicated than these is the familiar square knot, often

Figure 2.9 Mankind's two simplest knots are the overhand knot (right) with an axis of twofold rotational symmetry, and the figure-eight knot (left) with a center of inversion.

used to tie two cords end to end, of which a wooden model appears in fig. 2.10. In knotsman's language a knot serving this purpose is called a "bend." The square knot boasts both a plane of reflection symmetry and an axis of twofold rotational symmetry perpendicular to that plane, and therefore the point at which the axis meets the plane is a center of inversion symmetry.

It is interesting to turn from these knots, which are among man's oldest, to man's newest knot (fig. 2.11). On October 6, 1978, just before its one-year interregnum, the *Times* of London announced on its front page the discovery of a knot for tying two ropes end to end, promptly christened "Hunter's Bend" after its discoverer, Dr. Edward Hunter, a retired British physician.

In the words of its inventor, in order to tie it, "Lay the two ropes parallel with the ends opposite; throw a bight on the dou-

Figure 2.10 The square knot has a plane of reflection symmetry, horizontal in this picture, and an axis of twofold rotational symmetry perpendicular to the reflection plane.

Figure 2.11 Hunter's Bend, the recently invented compound of two overhand knots, has three mutually perpendicular axes of twofold rotational symmetry.

Figure 2.12 These wooden models show the courses of threads in a puzzle that asks which configuration is knotted. (The configuration at the right is knotted.)

ble (overlapping) part; tuck the ends through the bight in opposite directions, and a pull on the standing parts will draw it snug." Notice that Dr. Hunter's prescription follows almost exactly a description of how to tie an overhand knot in a single rope. Indeed Hunter's Bend is simply an ingenious and easily tied compound of two overhand knots. The compound has three mutually perpendicular axes of two-fold rotational symmetry. It forms a satisfactorily tight connection between any two of its four projecting ends.

The questions of when and whether a single line is knotted can be difficult to answer. Figure 2.12 shows a sculptural adaptation of a puzzle printed on a paper napkin asking which configuration is knotted. It may also be interesting to inquire whether any of the courses for automobiles in the interchanges of the preceding chapter is knotted. For an experimental approach to the answer, lay a string along the course and then pull on its two ends to see whether it straightens into an unknotted line.

3

GORDIAN KNOTS

Legend tells that Gordius, a peasant who became King of Phrygia, tied a knot of fiendish ingenuity out of cornel bark, fastening a wagon pole to a yoke. An oracle declared that whoever could untie it should reign over all Asia. Alexander the Great, unable to untie it, cut the knot with his sword, thus destroying forever that topological wonder. It is comforting to remember that Alexander's brutal act did not make him ruler of all Asia. And yet, even today, "to cut the Gordian Knot" remains an honorific expression.

For the past century and a half mathematicians have concerned themselves with what might well be called "Gordian knots"—knots that cannot be untied without being cut. The simplest of these knots is the trefoil knot (fig. 3.1). It follows the course of a continuous curve in three dimensions which joins upon itself, having neither beginning or end, but it cannot be turned into a simple loop without being cut. It can be looked upon as obtained by joining the two free ends of an overhand knot (fig. 2.5, right) in such a way that the joint is invisible.

The failed effort to untie them provides a method for classifying Gordian knots. Lay the knot flat and manipulate it until its parts cross one another a minimum number of times. The knot "belongs to" the class of that number. The trefoil knot is the only Gordian knot with just three crossings.

Figure 3.1 Connecting the free ends of an overhand knot (fig. 2.5) produces the simplest Gordian knot, the "trefoil knot."

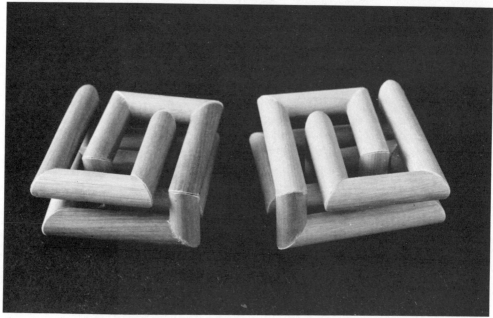

Figure 3.2 The trefoil knot can take either a left-handed or a right-handed form.

Figure 3.3 The sculptural appearance of the trefoil knot can exploit its axis of threefold rotational symmetry.

But the mirror image of the knot is distinguishable. A knot tied according to that image cannot be manipulated into identity with the original (fig. 3.2). Nevertheless in an accounting of Gordian knots the trefoil knot is ordinarily counted as a single knot, not two knots, and this convention will be adopted here.

Notice that even this simplest Gordian knot can take many different appearances. The sculptural forms shown in figs. 3.1–3.3 are all the same knot. A piece of string that follows the course of any one of them can be manipulated without cutting it so that it follows the course of any other.

The next more complicated Gordian knot arises when the ends of the figure-eight knot (fig. 3.5, left) are joined. The resulting knot has an irreducible minimum of four crossings (fig. 3.4). Among mathematicians it is famed as "Listing's Knot."* Again

*J. B. Listing, 1808–1882, a German mathematician who gave early notice to knots in 1847 and is sometimes regarded as the founder of knot theory.

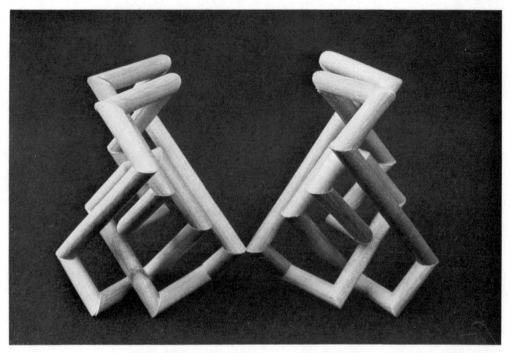

Figure 3.4 Listing's Knot results from connecting the free ends of the figure-eight knot. What appear to be distinguishable right- and left-handed forms are interconvertible without their having to be cut when they are made of flexible material.

there is only one knot in its class of four crossings, and again it can take many appearances (fig. 3.5). But it differs from the trefoil knot in failing to have right and left forms: it is "amphichiral."*

The move to Gordian knots with five crossings brings with it a new complication. Here are two distinguishable knots which are not mirror images of each other. The more obvious of them is the cinquefoil knot (fig. 3.6). It is especially attractive when it is laid out in a way to display an axis of fivefold rotational symmetry. It is interesting to compare this knot with the trefoil knot (fig. 3.1) and notice why a comparable tetrafoil knot does not exist.

*A word introduced by P. G. Tait toward the end of the last century to describe knots that can be deformed into their own mirror images.

Figure 3.5 Topologically these three versions of Listing's Knot are al!
alike: each has four irreducible crossings.

Figure 3.6 The cinquefoil knot is an especially handsome Gordian knot.

Figure 3.7 A knot with five crossings (left), distinct from the cinquefoil knot, has no axis of fivefold rotational symmetry, but only an axis of twofold rotational symmetry lying in the plane of the knot. In this respect it is comparable with a knot of six crossings (right).

When it is laid out flat for counting its crossings, the cinquefoil knot can be manipulated so that its five successive crossings fall on a single straight line. But the other knot of five crossings (fig. 3.7, left) cannot. Nor can it be manipulated into a form with an axis of fivefold rotational symmetry. Both species of knots can take distinguishable right- and left-handed forms.

When you connect the free ends of the familiar square knot so as to form a continuous path, you produce the first of the four species of Gordian knots with six crossings (fig. 3.8). Laid out flat it clearly has a plane of reflection symmetry perpendicular to its axis of figure and thus it cannot adopt right and left handed forms. The second species of knot with six crossings, obtained by joining the free ends of the granny knot, (Fig. 3.9, top) lacks any

Figure 3.8 Connecting the free ends of the familiar square knot constructs a knot of six crossings, with a plane of reflection symmetry perpendicular to its axis of figure.

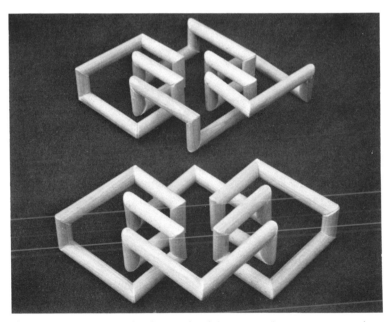

Figure 3.9 Connecting the free ends of the granny knot constructs a knot (above) with six crossings, without the plane of reflection symmetry of the square knot (below).

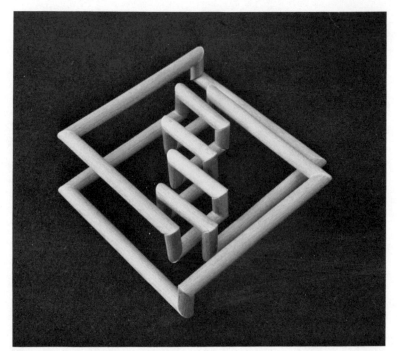

Figure 3.10 This is the fourth species of knot with six crossings.

plane of reflection and therefore appears in distinguishable right and left versions.

There are two other species of knots with six crossings, (fig. 3.7, right), the "Stevedore's knot," and (fig. 3.10). Indeed, as the number of crossings increases, the number of distinguishable species of knots increases quite rapidly. The familiar bowline knot fathers one of the many knots with eight crossings (fig. 3.11). One of the "Turk's Head" knots fathers another (fig. 3.12).

As the previous chapter showed, the newly invented knot, Hunter's Bend, is an interlock of two overhand knots. Hence the free ends of each knot can be connected to form an interlock of two trefoil knots. Alternatively the two overhand knots can be cross-connected to form a single Gordian knot. The three crossings of each overhand knot survive in the compound, and the interlocking entails six more, between the two knots. Thus Hunter's Bend fathers a Gordian knot with twelve crossings (fig. 3.13).

Figure 3.11 Connecting the free ends of the bowline knot constructs a knot with eight crossings.

Figure 3.12 This knot with eight crossings is produced by connecting the free ends of a Turk's head knot.

Figure 3.13 Connecting the free ends of Hunter's bend can produce a Gordian knot with twelve crossings.

The interlock of two trefoil knots suggests the versatility of the trefoil knot in forming composites. Thus interlocking four such knots produces the assembly shown in fig. 3.14. The directive principle employed in forming this assembly was the production of a structure with three mutually perpendicular axes of twofold symmetry.

It is interesting to notice in fig. 3.15 that there is a simple way to construct at least one knot with any chosen number of crossings: twist one rope around another and join the free ends of the first to those of the second. Notice also that the somewhat similar scheme of construction shown in fig. 16 succeeds in producing knots with any odd number of crossings but fails for the even numbers, providing instead two loops linked through each other.

The linkage is not simple. It has an irreducible number of crossings, like a Gordian knot. But unlike a Gordian knot the link-

Figure 3.14 This rigid interlock of four trefoil knots has three mutually perpendicular axes of twofold symmetry.

Figure 3.15 This scheme of construction can produce a knot with any number of crossings.

Figure 3.16 This scheme of construction will produce knots with odd numbers of crossings, but the even numbered constructions form linking loops.

age provides two paths which return to their starting points after traversing only half the structure. Thus it is not surprising to encounter structures which can be analyzed into two Gordian knots linked to each other.

Indeed simple rings are astonishingly versatile in their interconnections and interpenetrations. Even a single ring can afford delightfully various structures, as the next chapter shows.

Figure 3.17 Listing's Knot will provide a dramatic piece of constructivist sculpture.

4

CAT'S CRADLE

Viewed from a distance, two people playing Cat's Cradle with each other remind you of two deaf-mutes conversing, until you see that they hold between them a piece of string. In the game that they play, one holds a closed loop of string taut between his two hands, and the other inserts the fingers of his two hands symmetrically through the loop and releases it from his companion and pulls it taut in his turn, displaying a symmetrical figure of string. The first player then returns to perform similar operations on the loop, carrying it into a different design between his hands. Continuing, the players can traverse at least seven different designs before exhausting the possibilities accessible in this way.

Clearly Cat's Cradle has nothing to do with the beloved domestic animal after which it is named. To be sure, one station in its sequence is sometimes called Cat's Eye. But that name probably dates from relatively recent times, and it is called in question by the Korean name Cow's Eyeball. Indeed "cat's" is probably a corruption of the old English *cratch* (compare the French *crêche*) meaning *manger*.

The seven stations traversed in the sequence of play appear in fig. 4.8, with short sections of dowel representing the players' ten fingers. Since at each move a player inserts his fingers in two sides left free by his fellow, the axis of figure turns through a right

Figure 4.1 At the point where the original string for playing Cat's Cradle has been stretched into the first station, the only surviving symmetry of the ring is its center of inversion symmetry. In this sculptural adaptation of the station, the positions of the fingers have been displaced slightly in order to give the sculpture a pleasing appearance. Each such position is embraced by an interior angle, and all such angles represent finger positions. The only exceptions are at the very short pieces, enabling the structure to maintain continuity as it moves from one level to another. This station is named Cradle by English speakers, but in Korea it has the less endearing name "cover for a hearse."

angle between successive stations. The moves of the players are clearly described by Jayne.*

Because the taut strings form figures composed of straight lines, they make models for constructivist sculpture requiring little adaptation. The models appearing in figs. 4.1–4.7 carefully retain

*Carolyn Furness Jayne, *String Figures and How To Make Them* (New York: Dover, 1962).

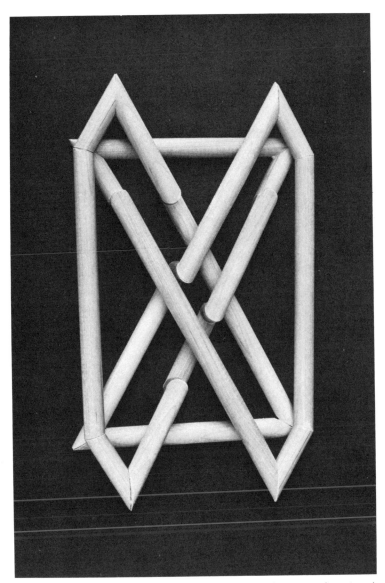

Figure 4.2 The manipulation of the string into the second station has destroyed all its symmetry. The resulting figure has been decorated by a wide variety of names. In the United States it is known as either Soldier's Bed or Fish Pond. In Korea it is called Chess Board, and in England, Church Window. The Japanese call it a Mountain Cat!

CAT'S CRADLE

Figure 4.3 The manipulation yielding the third station has constructed a pleasingly high degree of symmetry. A plane of reflection symmetry passes perpendicular to the figure, and perpendicular to that plane is an axis of twofold rotational symmetry. As a mathematically necessary result of this combination of axis and plane, the figure has a center of inversion symmetry where the axis meets the plane. By various stretches of the imagination this station has received the name Candles in the United States, Chopsticks in Korea, Koto (a musical instrument) in Japan, and Mirror in Denmark.

Figure 4.4 The station known as Manger in the United States seems to be relatively free of other names in other countries. But its resemblance to the first station has led to the name of Inverted Cradle in England. Again the structure has a center of inversion symmetry and no other.

Figure 4.5 In the fifth station the ring of string has constructed the mirror image of Soldier's Bed. In the United States this structure is called Diamonds, in England Soldier's Bed Inverted.

the crossings of the string figures. In order to do so, the wooden members must cross one another at different levels, as in knots of the preceding chapter. Usually two distinguishable levels suffice to accommodate the crossings, but in the later figures of the succession three levels are necessary. The increase in thickness

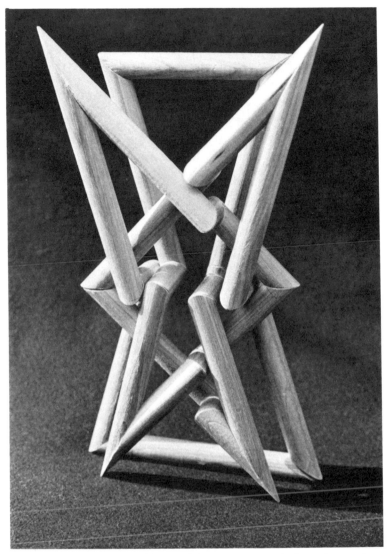

Figure 4.6 The English hold over the name Diamonds from the previous station and apply it to this. But in the United States it is called Cat's Eye, in Korea Cow's Eyeball, and in Japan Horse Eye. Here the wooden models can no longer display the crossings correctly if they are confined to two levels. The addition of one more level also enables the model to stand erect, forming a dramatic statue without rigorous symmetry, but with an agreeably symmetrical appearance.

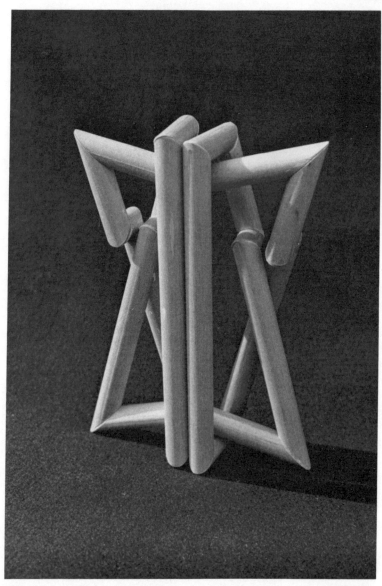

Figure 4.7 A sculptural model of the seventh station occupies planes at three levels and again it will stand erect monumentally. When it stands so, its symmetry is similar to that of a human being: one vertical plane of reflection. Nevertheless in the United States it is called Fish-in-a-Dish, in Korea Rice-Mill Pestle and in Japan Tsuluni (a musical instrument).

Figure 4.8

front-to-back thus entailed provides sufficient stability to enable these models to stand with their axes of figure vertical.

Evidently all the stations of Cat's Cradle are unknotted rings. As soon as a player removes his hands from it, any configuration falls into a ring. But this may seem a questionable test, if by accident some part of the cord catches another part so as to make a tightly held loop. Analogous loops which would offer similar difficulties of classification can be easily imagined. Consider, for example, what would be made by tying a ring of string in an overhand knot. Clearly we would call the result a knot. But clearly we could convert the result into a ring again without cutting it. The resolution of this and other such difficulties demands subtler analyses than this book can offer.

5

CHAINED POLYLINKS

Polylink seems a suitable name for an assembly of rings in which each ring is flat, and built of straight-line segments, and the rings are linked to one another. When the rings are linked in succession, they form the simple structure usually called a chain.

The chain gains new interest when its two ends are linked through each other to form a ring of rings (fig. 5.1). Especially interesting examples arise when the component rings are shaped as equilateral triangles and the ratio of length of each side to the diameter of the rods is approximately six. Then the ring of rings often can be arranged into a rigid assembly.

The resulting assembly of rings has lost the appearance of a chain, even though nothing has occurred to alter the fact that each ring is linked through only two others in a chained succession. Although its rings are shaped as regular polygons, the structure is not "regular," however, because its corners fall into several distinguishable classes.

A remarkably simple rigid chained polylink is formed by four equilateral triangular rings (fig. 5.2). In order to assure yourself that this structure is indeed a chain, start numbering the component rings one by one. Each successive ring should form a link with its predecessor.

The twelve corners of this structure fall into two distinguisha-

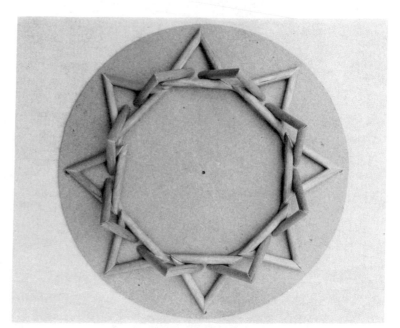

Figure 5.1 Chaining sixteen equilateral triangular rings and finally linking the opposite ends of the chain forms a loose ring of rings.

ble classes. Four come in pairs at opposite ends of the axis of figure, and eight come in four pairs in an equatorial band.

Notice in fig. 5.2 that the completed polylink can be regarded as gathered from its parental ring of rings in two stages. In the first stage two alternate rings are brought together; and in the second stage the remaining two rings are brought together in such a way that a pair of their corners points oppositely to a pair of corners of the previously gathered rings.

Chained polylinks of similar rigidity arise from rings of rings with more members. Their symmetry is greatest when an even number of component rings participates. Figures 5.3–5.5 show the products of rings of six, eight and ten chained equilateral triangular rings. First the alternate rings are gathered together into a pyramid, and then the remaining rings are similarly gathered into a pyramid pointing in the opposite direction.

It is worth remarking that the final assemblies of chained rings are quite symmetrical. An assembly of this type with $2n$ constit-

Figure 5.2 The ring of four equilateral triangular rings acquires a pyramid (left) when alternate rings are gathered. It acquires a second pyramid, and rigidity, when the remaining rings are gathered (right).

Figure 5.3 In the ring of six rings, gathering the first pyramid (left) makes the threefold rotational symmetry of the assembly conspicuous. The two pyramids (right) are tightly knit.

Figure 5.4 The ring of eight chained rings becomes quite octahedral in appearance when the rings are gathered into two pyramids.

Figure 5.5 The pentagonal ring of ten equilateral triangular rings retains a pleasing symmetrical appearance when either five (left) or ten (right) of its rings are gathered.

uent rings has an axis of n-fold rotational symmetry passing through the peaks of its two pyramids, and n axes of 2-fold rotational symmetry perpendicular to that principal axis. Since none of these assemblies has either a plane of reflection or a center of inversion, however, all can take distinguishable right-handed or left-handed forms.

Such chained polylinks as those shown in figs. 5.3–5.5 make a direct connection with the paths known to mathematicians as "Hamilton circuits." A century ago Sir William Rowan Hamilton proposed an interesting pursuit with convex polyhedra. Regard one such polyhedron as representing the surface of the earth, and each of its corners as representing a city. The edges of the polyhedron now represent roads between those cities. Sir William suggested that you can find routes along those roads which enable you to visit each city just once and finally to return to your starting place.

Evidently the rings of rings portray Hamilton circuits if you correlate each ring with a vertex of a polyhedron and each linkage with an edge. In this way any closed chained polylink made of equilateral triangular rings could be correlated with a semi-regular antiprism.* But the appearances of polylink and antiprism may be deceptively different. The visual imagination has difficulty shrinking a ring to a point, and the natural visual tendency to correlate each side of a ring with the edge of a polyhedron is frustrated when two rings link the same side of a third ring.

In principle chained polylinks could be made of rings with any polygonal shape. Those described here, made of equilateral triangular rings, fall in a subclass of polylinks that might be appropriately called deltalinks.

Solid geometers often single out for special attention a species called "deltahedra." These are solids with flat faces shaped as equilateral triangles. They derive their name from the fact that the equilateral triangle is the shape of the Greek capital letter delta. The shape is uniquely versatile in proliferating rugged and interesting structures, not only in its use in polyhedra, but also in polylinks.

The stability and rigidity of a deltalink are improved still further by abandoning the requirement that it be chained. Thus, for example, the most obvious way to stiffen almost any deltalink is to cross-link it to other parts of itself. Thereupon certain rings acquire more than the two chaining links.

Clearly the rigidity of a polylink can be expected to improve

*Shapes, Space, and Symmetry, p. 61.

Figure 5.6 Three equilateral triangular rings forming a Borromean configuration can be held in rigidity by a fourth identical ring to form a "nolink."

with the addition of more links restraining the relative motion of its rings. But there is at least one polylink that rigidly maintains its shape with no links whatever, the "nolink" shown in fig. 5.6. In this assembly of four rings, three have the configuration of the "Borromean rings"* and the fourth props the three apart so that they cannot fall toward one another. It is wholly unexpected that there should exist a configuration of unlinked rings with such rigidity.

The idea of cross-linking the rings of a chain focusses attention on polylinks in which each ring links with three others, instead of two. Among deltalinks it is tempting to nominate the four linking rings, all centered at the same place (fig. 5.7) as the simplest.† But there is another way of assembling four deltarings triply linked that is equally simple (fig. 5.8)§ Notice that both of these assemblies, simple though they are, come in, right-handed and left-handed forms.

*Ibid., p. 183.
†Ibid., p. 182.
§This assembly was first pointed out to the writer by Morton Matthew.

Figure 5.7 Four equilateral triangular rings can link with one another to form either a right-handed or left-handed assembly with the rings centered at the same point.

Figure 5.8 Four equilateral triangular rings can form a regular polylink without centering all the rings at the same point.

Notice also that in the last two clusters of four deltarings, you have encountered for the first time examples of regular polylinks. The idea of a regular polylink is analogous to that of a regular polyhedron.§ Such a polylink obeys two requirements: (1) all its rings are shaped as congruent regular polygons, and (2) all corners are identically surrounded in the structure. It has been proved that there are only nine polyhedra fulfilling these requirements. But there are many more regular polylinks. The rest of this book is devoted to describing some.

*Shapes, pp. 1, 70.

6

REGULAR POLYLINKS, 1

To require that a polylink be "regular"—made of rings shaped as regular polygons whose corners are indistinguishable in the structure—is to delimit its structure more stringently than might be imagined offhand. For example, each ring must stand perpendicular to an axis of rotational symmetry with the same count as that of its corners, and all rings must have the same size and must lie at the same perpendicular distance from a single identifiable center of the structure.

These requirements strongly suggest the use of the regular polyhedra* as guides to the construction of regular polylinks. Rings of the appropriate shape and size can be laid against the faces of a polyhedron. They will then automatically be perpendicular to appropriate axes of symmetry and at the same distance from the center of the structure.

The rings can then be turned in their own planes about their centers in the same sense and to the same extent, clockwise or counterclockwise, until they do not clash, but link with one another. The two senses of rotation provide two alternative polylinks, both regular, which are mirror images of each other. Figure 6.1 shows successive steps in the procedure just described; figure 6.2 shows the two products.

*Holden, *Shapes, Space, and Symmetry.*

Figure 6.1 One way to produce a threading diagram for a regular poly-link is to link cardboard rings on a cardboard polyhedron.

Figure 6.2 The procedure of fig. 6.1 produces two enantiomorphous regular polylinks.

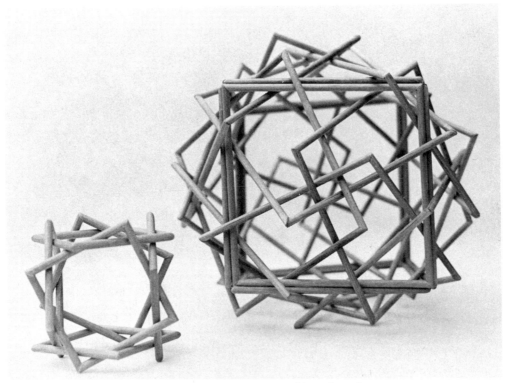

Figure 6.3 Three examples of a cubic model (left) suitably enlarged can interpenetrate to form a structure that is again cubic, but is not regular.

It is interesting to notice in passing that a delightful polylink can be made (fig. 6.3) in which three of these cubical structures interpenetrate one another with their centers at a common point. The directive idea lying behind this compound is to retain in the result the same symmetry as that possessed by each of the three ingredients: three axes of fourfold, four axes of threefold, and six axes of twofold rotational symmetry. Notice that, quite unexpectedly, the compound has edges along which the sides of the rings lie in parallel pairs outlining a cube. But the compound is not regular because its corners fall in three distinguishable groups. Twenty-four corners form eight groups of three at the corners of the outlined cube (fig. 6.4), eight more groups of three surround those groups at some distance, and six groups of four fall near the centers of the faces of the outlined cube (fig. 6.3).

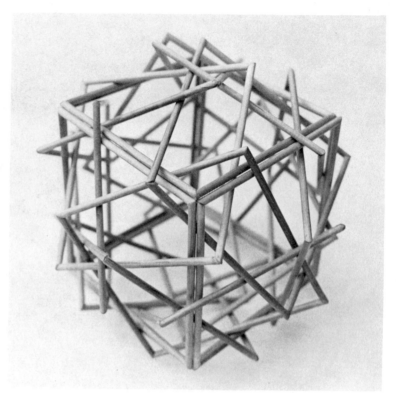

Figure 6.4 The corners of the compound of three cubic structures (fig. 6.3) fall in three distinguishable groups.

Another structure (fig. 6.5) made of square rings, again with the symmetry of a cube but again not regular, flows from the familiar rhombic dodecahedron. A rhombus, with two pairs of parallel sides, all of equal length, is an inviting surrogate for a square, and the rhombic solids can be used to direct the arrangement of square rings. But the solids faced by them are at best the duals of semi-regular solids (*Shapes, Space, and Symmetry*, p. 50), and thus the polylinks flowing from them are not regular.

Anyone in hot pursuit of regular linkages of square rings will finally run down the structure shown in fig. 6.6. This structure is not directly accessible by the procedure pictured in fig. 6.1, but only by a generalization of the ideas outlined there. The rings are too large to rest on the faces of a cube; they will rest upon

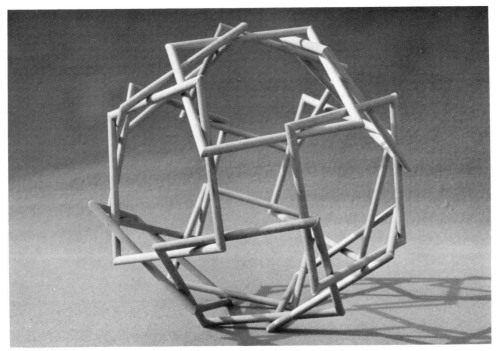

Figure 6.5 Twelve square rings lying parallel to the faces of a rhombic dodecahedron form a polylink with cubic symmetry that is not regular.

Figure 6.6 In a second regular cubic polylink, the rings of each parallel pair are closer together than those of fig. 6.2.

Figure 6.7 Regular hexagonal rings can form a structure in which rigidity results from the fact that each ring embraces its companions instead of linking with them.

extensions of those faces. Cardboard arms extending from the cube's faces will support "threading diagrams" analogous to the diagrams of fig. 6.1. Clearly this device can insure that the linked rings are suitably positioned, oriented, and distanced from a common center.

A curious little array of rings arises from placing three regular hexagonal rings so that they lie in mutually perpendicular planes and define a common center (fig. 6.7). All the corners fall on a common sphere. But there are two distinguishable groups. One group comprises the six corners that lie snugly above the sides

Figure 6.8 The sides of the hexagons of fig. 6.7 can be extended to form "Stars of David."

of the component rings and thus bind the whole structure into rigidity. The remaining twelve corners float freely. Stellation of the regular hexagons in the structure shown in fig. 6.7 converts them into three "Stars of David" and hence into a star-hexagonal array, fig. 6.8.

Before looking further at rings that are not square, it is rewarding to examine some structures flowing from the Archimedean semiregular solids. Figure 6.5 showed such a polylink, of course, since the rhombic dodecahedron is the solid which is dual to the Archimedean cuboctahedron. Another Archimedean solid is the small rhombic cuboctahedron whose faces comprise eighteen squares and eight triangles. The polylink flowing from it appears

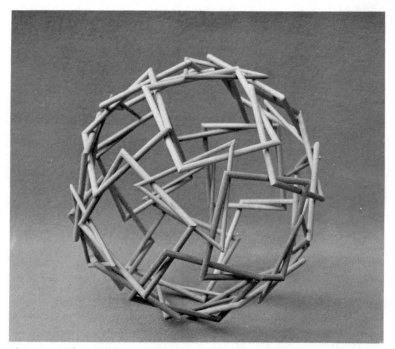

Figure 6.9 This polylink, with eighteen square and eight triangular rings, is built upon an archimedean semiregular solid, the small rhombic cuboctahedron.

in fig. 6.9. Since its parental polyhedron is not regular there is no reason to believe that the corners of the polylink should all lie on the surface of a single sphere. But the 96 corners do give the polylink a remarkably spherical appearance.

In a similar way the corners of the polylinks fathered by the snub cube, whose corners are still more numerous, 120, give even closer approximations for a spherical appearance (fig. 6.10). In contemplating their use as bases for polylinks, the snub cube and the snub dodecahedron raise an interesting question. These are unique among the semi-regular solids in their trait of appearing in enantiomorphous forms. Experience so far would predict that each of these forms would separately lead to two enantiomorphous polylinks. Figure 6.11 shows that this expectation is fulfilled; the Archimedean snub cube fathers four distinguishable polylinks. Whoever would make models of these polylinks will find the appended diagrams helpful.

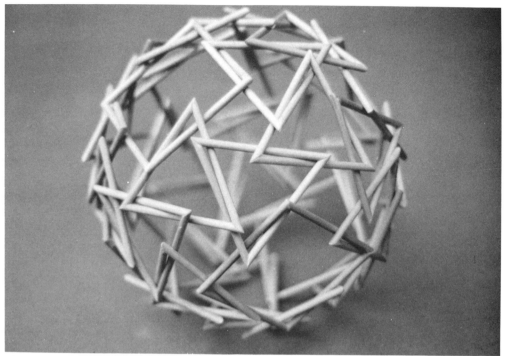

Figure 6.10 The polylinks fathered by the Archimedean snub cube have six square and thirty-two triangular rings.

Figure 6.11 The two enantiomorphous Archimedean snub cubes father four distinguishable polylinks.

7

REGULAR POLYLINKS, 2

To most people squares will seem to be more natural inhabitants of space than regular pentagons. Placed side by side against one another, regular pentagons cannot fill a plane completely, as squares can. Indeed it always seems surprising that twelve regular pentagons meeting side by side can make an exactly closed figure.

But in fact the resulting regular dodecahedron furnishes one of the five regular solids that form such convenient sources of regular polylinks (fig. 7.1).

A procedure entirely analogous to that outlined for squares in the beginning of the previous chapter produces two regular enantiomorphous polylinks in which each pentagonal ring links with five surrounding it (fig. 7.2).

Notice now that there are a few rules which pentagonal rings must obey in order to form a regular polylink. The rings must lie in pairs on parallel planes, with each pair perpendicular to an axis of fivefold symmetry. Each ring must link with five or some multiple of five other rings.

In searching for a polylink in which each ring links with ten others instead of just five it is necessary to use extensions of the dodecahedral faces. A relatively large structure (fig. 7.3) in which each ring links ten others is a most arresting product when made

Figure 7.1 The regular pentagonal dodecahedron forms the basis for threading diagrams for regular pentagonal polylinks.

of ¼″ doweling because of its size. Its sixty corners appear in twelve groups of five over the centers of the twelve rings.

Placed side by side with the polylink shown in fig. 7.2, the polylink in fig. 7.3 is very much larger. But this difference is an illusion of scale. The structures pictured are all made of cylindrical rods whose diameters are one quarter inch. Any can be scaled upward or downward linearly by changing the diameter of the doweling used, as fig. 7.4 shows. If the structure shown in fig. 7.2 were made of ½″ doweling, it would top the structure shown in fig. 7.3 by three inches.

Another regular polylink with pentagonal rings, each linking with only five others, appears in fig. 7.5. This polylink differs from that in fig. 7.2, with the same linkage number, by appearing no-

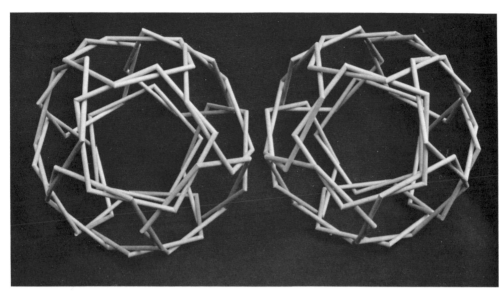

Figure 7.2 The embodiments of threading diagrams, constructed as fig. 7.1 shows, can take either a right-handed or a left-handed form.

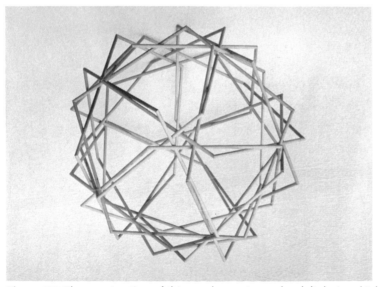

Figure 7.3 The construction of this regular pentagonal polylink, in which each ring links with ten others, is clearest when it is viewed along an axis of fivefold rotational symmetry.

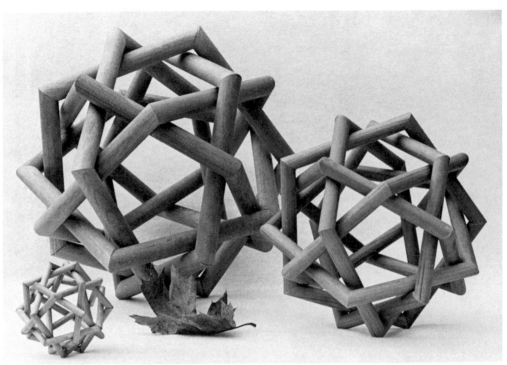

Figure 7.4 These three polylinks, of widely different sizes, have the same structure.

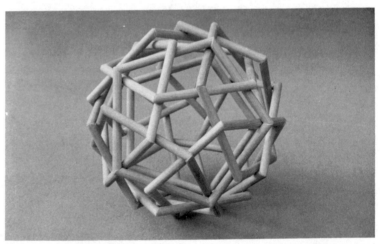

Figure 7.5 In this regular polylink, as in that pictured in fig. 7.2, each ring links with five others (see appendix).

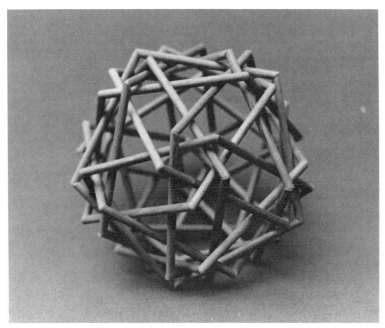

Figure 7.6 In this polylink, as in that of fig. 7.3, each ring links with ten others (see Appendix).

tably smaller. It accomplishes this contraction by pulling closer together the parallel planes on which the rings lie, and permitting more of the substance of each ring to penetrate the planes of the others. Five projecting corners provide the outline of a small pentagon within each ring.

When the parallel planes move still closer together, each ring must link with ten others, as in the structure shown in fig. 7.6. The more complicated threading entailed by this linkage appears in the accompanying diagram. Again five projecting corners outline a little pentagon within each ring.

When the parallel planes of each pair are made to coincide, their two pentagonal rings lie against each other (fig. 7.7). Here the dodecahedron, to whose faces those planes are parallel, has shrunk to the vanishing point. Each ring links with the ten rings that are admissible to it; that is to say with all of the twelve rings except itself and the ring parallel to it.

The doubling of the rings in this structure immediately suggests

Figure 7.7 In this regular pentagonal polylink, the rings of each parallel pair lie against each other.

congealing each pair of rings into a single ring. In the resulting polylink (fig. 7.8) each of the six rings links with all its five companions, and the structure is again regular. All six rings hold their centers in common. This is the polylink used to demonstrate scaling in fig. 7.4.

Since the time of Johannes Kepler (1571–1630) the celebrated astronomer, geometers have recognized the importance of a class of regular polygons supplementing the more familiar convex polygons. Like them the newer polygons have equal angles at their corners, and their sides are all of equal length. But unlike them, the sides cross one another. They are often called "star-polygons."

The accompanying diagram (fig. 7.9) makes clear the nature of the star-pentagon (left) and of the problem that it presents to the construction of a regular star-pentagonal polylink. Often the polygon is thought about in the simplified form shown at the center

Figure 7.8 In the regular polylink made of six pentagonal rings, the rings all have a common center and each is equatorial.

Figure 7.9 Since the star-pentagon (left), sometimes called the "pentagram," is regular, it can father a ring suitable for a regular polylink. The simplified form (center) is more useful than the rigorous form (right).

of the diagram. If the polygonal rings are constructed as shown at the right, they can be linked in two distinguishable ways: through either the triangular regions or through the pentagonal region. For the construction shown here, this purism has been ignored and the simplified form of the star-polygon has been used.

The arrangement of pentagons in the structure shown in fig. 7.8 has directed the arrangement of star-pentagons in the structure shown in fig. 7.10. In an analogous way the pentagonal rings

Figure 7.10 A regular polylink with star-pentagonal rings can be based on the structure shown in fig. 7.8.

in the structures pictured earlier in this chapter can direct the construction of other regular star-pentagonal polylinks. Thus the total number of regular polylinks based on the regular pentagon is much larger than the number based on the square (chapter 6). But it is still not as large as the number based on the triangle, as the next chapter will show.

8

REGULAR POLYLINKS, 3

The classical regular polyhedra include three whose faces are shaped as equilateral triangles, the tetrahedron, the octahedron, and the icosahedron. The two regular polylinks based on the tetrahedron have already been pictured in figs. 5.7 and 5.8. The four rings composing the polylink of fig. 5.7 are concentric and equatorial like the six pentagonal rings of the structure shown in figs. 7.8 and 7.10.

In the second tetralink the four triangular rings take less simple courses. But, arrived at by the procedure using a threading diagram on the faces of a tetrahedron, the structures clearly embrace a tetrahedron. Several structures shown later in this chapter can be regarded as compounds in which more than one of these tetrahedral structures interpenetrate.

The use of a threading diagram on the faces of a regular octahedron leads to the simplest structure with eight rings (fig. 8.1) in which each ring links with three others. Viewed along the axes of fourfold rotational symmetry, these structures strongly suggest the close relation between the octahedron and the cube. The triangles in each parallel pair are perpendicular to the body-diagonals of a cube.

When the parallel planes on which the rings lie are brought closer together, each ring is able to link with six others, as fig. 8.2 shows. Here is the first of the regular polylinks that can be

Figure 8.1 From threading diagrams on a regular octahedron, two enantiomorphous regular polylinks appear, in which each of the eight rings links with three others.

Figure 8.2 An octalink in which each ring links with six others is viewed here along an axis of twofold rotational symmetry.

Figure 8.3 The fact that the octalink shown in fig. 8.2 can be regarded as a compound of two regular tetralinks of the type shown in fig. 5.8 is dramatized by making the component tetralinks of two different colors.

regarded as a composite of two others. Figure 8.3 shows how the two enantiomorphous forms are compounded from the two corresponding regular tetralinks shown in fig. 5.8.

Finally, when the parallel planes are brought even closer together, the corresponding rings come into contact, producing the structures shown in fig. 8.4. As in the case of the comparable pentagonal rings (fig. 7.7), the rings can be fused in pairs to produce a regular polylink with half as many rings. This tetralink has already been shown in fig. 5.7, a structure whose four rings are concentric and circumferential.

Regular polylinks built upon the regular icosahedron are more numerous than those based upon any of the other four regular polyhedra. Until now twelve such polylinks have been discovered, and no doubt more will follow. The twenty faces of the icosahedron come in parallel pairs perpendicular to the axes of

Figure 8.4 In this octalink the rings of the tetralink shown in fig. 5.7 are doubled.

threefold rotational symmetry. Each ring of any regular polylink must link with three, six, nine, twelve, fifteen, or eighteen other rings.

The complications which arise in threading many of these polylinks are sufficient to require a threading base which offers extensions of the faces of an icosahedron.* In fig. 8.5, for example, a stellation of the regular icosahedron is serving this purpose. For the first structure (fig. 8.6) the icosahedron itself suffices for a base. Here each of the twenty equilateral rings links with three others and the product embraces the icosahedron snugly.

Equally ready to hand is the simpler of the two structures in which each ring links with six others (fig. 8.7). The rings arrange

*See Magnus J. Wenninger, *Polyhedron Models* (Cambridge: Cambridge University Press, 1970), p. 55.

REGULAR
POLYLINKS, 3

74

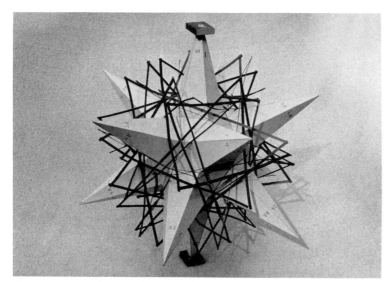

Figure 8.5 A stellation of the regular icosahedron forms a suitable base for threading diagrams of regular icosalinks.

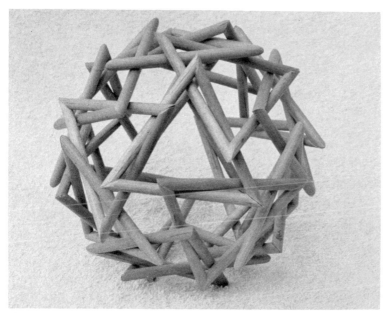

Figure 8.6 In the simplest of the many regular icosalinks each of the twenty equilateral triangular rings links with three others.

Figure 8.7 Five sets of four triangular rings, arranged like the faces of five interpenetrating regular tetrahedra, form a regular icosalink in which each ring links with six others.

themselves so that pairs of adjacent sides combine to outline the edges of five interpenetrating regular tetrahedra. A compound of five interpenetrating tetrahedra is a familiar stellation of the regular icosahedron (*Shapes, Space*, p. 39).

Unexpectedly there is a much larger self-supporting structure in which again each of twenty equilateral triangles links with six others (fig. 8.8). It stands more than twice as high as its predecessor, but it displays its detailed symmetries much less vividly. The resulting structure is the first of many regular icosalinks that can be regarded as compounds of five interpenetrating tetralinks of the type shown in fig. 5.7.

Much more closely related visually to the structure of fig. 8.7 is that shown in fig. 8.9, in which each ring links with nine others. Here the close bunching of corners in triples, which charac-

Figure 8.8 In this regular icosalink, the fact that each ring links with six others is obscure, nor is it obvious that the polylink is a compound of five interpenetrating tetralinks.

terized the former structure, is replaced by a close linkage of the corresponding triples. The added linkages enlarge the structure to more than one-and-one-half times its previous size. It forms the second example of a compound of five regular tetralinks.

There are at least three more regular polylinks in which each ring links with nine others. The polylink shown in fig. 8.10 is especially attractive. The corners of the twenty triangular rings form twelve little five-petalled rosettes projecting from the major

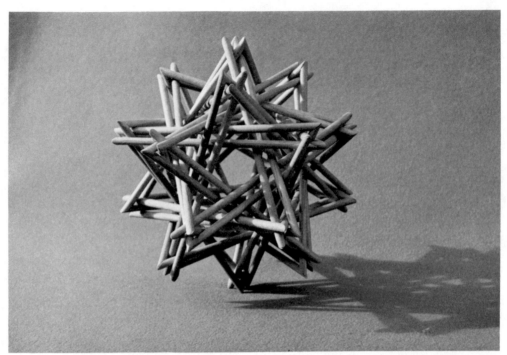

Figure 8.9 In this regular icosalink each ring links with nine others, and it is easy to see that the assembly is a compound of five regular tetralinks.

surface of the polylink. The rosettes are joined to one another by pairs of roughly parallel sides.

On the structure shown in fig. 8.11 the corners are quite evenly distributed, giving to the polylink a roughly spherical outline. On a smaller polylink pictured in fig. 8.12 the bunched corners border dishes with fivefold rotational symmetry and with regular pentagons outlined at the bottom of each. Both these polylinks can be compounded of five regular tetralinks.

Superficially the regular polylinks shown in fig. 8.13 and 8.14, in which each ring links with twelve others, look remarkably like the arrangement of rings shown in fig. 8.11. The polylink pictured in fig. 8.13 is that for which the threading diagram was shown in fig. 8.5. Again the corners of the rings outline dishes with fivefold rotational symmetry. In the polylink of fig. 8.14 the exterior is a

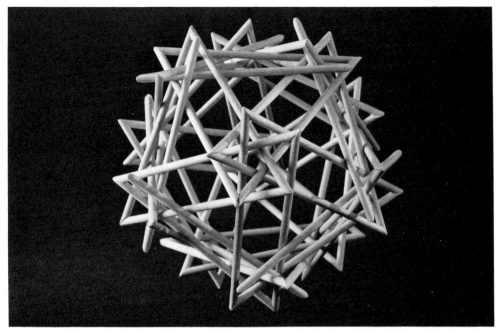

Figure 8.10 In this attractive regular icosalink, each ring links with nine others.

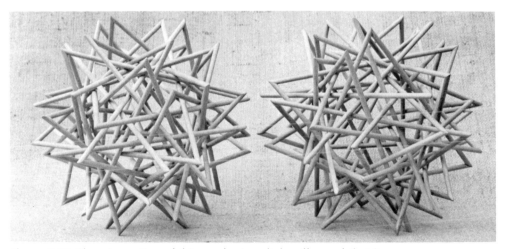

Figure 8.11 Close inspection of this regular icosalink will reveal that its rings, of which each links with nine others, form a compound of five regular tetralinks.

Figure 8.12 Superficially this regular icosalink looks somewhat like that shown in fig. 8.9; but its linkages, again nine per ring and resolvable into five regular tetralinks, are quite different in detail.

little less conspicuously dished and the threefold bunching of the corners is more prominent. Both can be regarded as compounds of five regular tetralinks.

Bringing closer together the planes of each parallel pair reveals at least three regular polylinks in which each ring links with eighteen others.

On the structure shown in fig. 8.15 the exterior dishes have threefold rotational symmetry and the corners are bunched in fives. The rings of each parallel pair stand a short distance apart, propped by other intervening rings. This regular polylink is again a compound of five regular tetralinks.

A still closer approach by the parallel planes brings to notice two different structures in which the members of a pair of rings

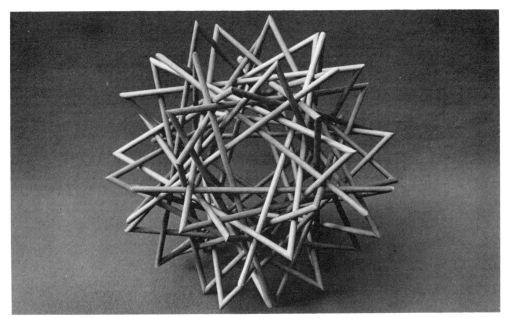

Figure 8.13 This regular icosalink, in which each ring links with twelve others, embodies the threading diagram of fig. 8.5, and is a compound of five regular tetralinks.

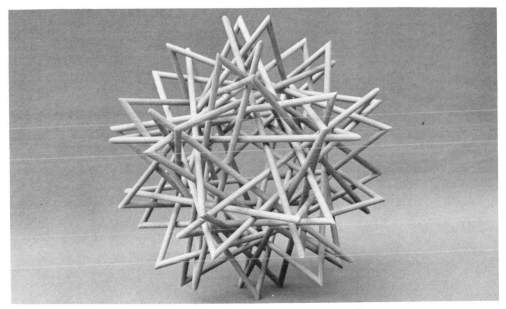

Figure 8.14 The icosalink pictured here is again a compound of five regular tetralinks; superficially very like that pictured in fig. 8.13. Though somewhat smaller, again each ring links with twelve others.

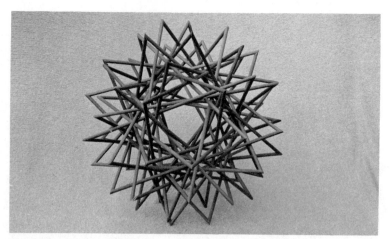

Figure 8.15 This regular icosalink is unusual in presenting dishes with threefold instead of fivefold rotational symmetry. It is a compound of five tetralinks and each ring links with eighteen others.

Figure 8.16 In this regular icosalink, a compound of five regular tetralinks, where each ring links with eighteen others, the rings lie against each other in pairs and are turned about each other in their common plane.

Figure 8.17 In their final approach to each other the twenty equilateral triangular rings form a regular icosalink dominated by the pairing of the rings.

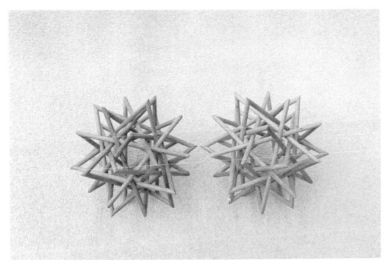

Figure 8.18 When the rings of fig. 8.17 coalesce in pairs, they form a regular icosalink of ten rings, concentric and equatorial, in which each ring links all its nine companions.

come into contact. In one structure, shown in fig. 8.16, the members of each pair of rings are turned in their common plane, admitting intervention by other rings. This structure is the last of the icosalinks compoundable by regular tetralinks.

The other structure, shown in fig. 8.17, achieves a final compression in which the rings lie congruently in pairs against each other. As with the earlier congruent juxtapositions, the structure shown in fig. 8.17 invites the coalescence of its parallel parts into a regular polylink with half as many rings, concentric and equatorial. The three regular polylinks, with four, six, and ten concentric and equatorial rings, appear in figs. 5.7, 7.8, and 8.18. They form the crown of the regular polylinks. Indeed they are the sovereign orderly tangles.

APPENDIX

Most of the models pictured in this book are made of ¼″ wooden doweling. Those in chapter 1 are made of #18 AWG wire. In the models of screens, knots, and cat's cradle, the lengths of rod employed are not critical: many are made of ½″ doweling and a few of ⅝″ or 1″ doweling.

For the regular polylinks and their compounds, the ratio of length to diameter is critical and the structures can be scaled up and down linearly in size. The following table lists the critical data applying to these polylinks when made of ¼″ diameter doweling. The "size" is the maximum diametral dimension of the model. Lengths are measured from tip to tip.

The following diagrams will help in making the four polylinks shown in figs. 6.10, 7.5, and 7.6.

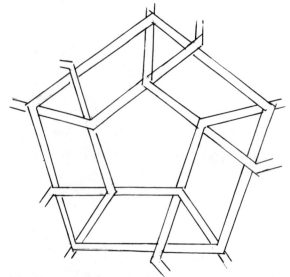

Figure A1.1

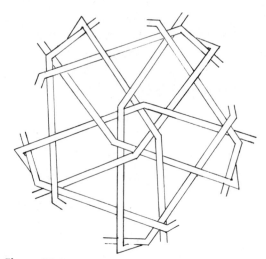

Figure A1.2

Figure A1.3

Figure A1.4

Figure A1.5

Figure A1.6

APPENDIX

90

This table provides the information necessary for replicating the polylinks shown in chapters 5–8.

Fig.	No. of Rings	No. Rings Linking	Length of Piece	Size
Fig. 5.6	4	0	3"	3½"
5.7	4	3	2¼"	2⅝"
5.8	4	3	3¼"	4¼"
6.2	6	4	3½"	6"
6.3	18	8 and 10	7"	12½"
6.5	12	3 and 4	4¾"	9"
6.6	6	4	2⅛"	3¼"
6.7	3	0	2¼"	4⅜"
6.8	3	0	1"	3¾"
6.9	18sq.8tr.	4 and 3	4¼"	11½"
6.10	6sq.32tr.	4 and 3	4¾"	11¾"
7.2	12	5	4½"	10½"
7.3	12	10	10"	18"
7.5	12	5	3¼"	$5^{13}/_{16}$"
7.6	12	10	4½"	7⅛"
7.7	12	10	3¼"	4¾"
7.8	6	5	2¼"	3¾"
7.9	6	5	1⅝"	4¼"
8.1	8	3	$3^{1}/_{16}$"	4¼"
8.2	8	6	4¼"	5⅛"
8.4	8	6	3¾"	3¾"
8.6	20	3	$3^{9}/_{16}$"	6⅜"
8.7	20	6	4¾"	5½"
8.8	20	6	10⅛"	$12^{1}/_{16}$"
8.9	20	9	7½"	8⅞"
8.10	20	9	7¼"	10"
8.11	20	9	10"	10⅞"
8.12	20	9	7¼"	8⅞"
8.13	20	12	11¾"	12⅝"
8.14	20	12	10	11⅝"
8.15	20	18	13⅝"	14⅞"
8.16	20	18	9½"	10¼"
8.17	20	18	8½"	9¾"
8.18	10	9	6⅜"	7⅜"

BIBLIOGRAPHY

The best source for a conspectus of plans for highway interchanges is *Traffic Engineering Handbook,* 3rd edition, Chapter 17. Inst. of Traffic Engineers, Wash., D.C. 1965.

A handsome book describing elementary weaving practice is *Spiders' Games,* a Book for Beginning Weavers by Phylis Morrison, University of Washington Press, Seattle, 1979.

A beautiful description of practical knots is provided by *The Ashley Book of Knots* by Clifford W. Ashley, Doubleday, Garden City, N.Y. 1944.

A good source for the classical theory of knots is *Introduction to Knot Theory,* Crowell and Fox., Ginn & Co., Boston 1963. A more modern treatment of knots is to be found in an article on "The Theory of Knots" by Lee Neuwirth, *Scientific Amer.* Vol. 240, No. 6, pages 110–124, N.Y. June 1979.

The classic treatment of symmetry by a master of mathematics is *Symmetry* by Hermann Weyl, Princeton University Press, 1952.

Cat's cradle and many similar configurations of a single ring are described in *String Figures and How to Make Them* by Caroline Furness Jayne, Dover Publications, Inc., N.Y. 1962.

The polyhedra, regular and irregular, referred to in this book are described and pictured in *Shapes, Space and Symmetry,* by Alan Holden, Columbia University Press, N.Y. 1971.

The original discussion of regular polylinks is by Alan Holden, *Structural Topology,* #4, pages 41–45, Montreal, 1980.

INDEX

Alexander the Great, 21
Amphichiral knot: *see* Knots
Antiprism, semi-regular, 49
Axes of rotational symmetry: *see* Symmetry

Basketry, 9
Bend: *see* Knots
Borromean rings, 50
Bowline knot: *see* Knots

Cinquefoil knot: *see* Knots
Classifying knots: *see* Knots
Cloverleaf, 2; nonredundant, 3
Compounds: knots, 19, 28, 30; polylinks, in cubical, 55; octalinks as, 73; tetralinks in, 73, 76, 77, 78, 79, 80, 81, 84; icosalinks as, 76
Connection between highways, 2
Cube, symmetry of, 13, 14
Cubical regular polylinks, 53; Compound of three, 55; second type, 56

Deltalinks, definition: 49

Enantiomorphs: in weaving, 13; as mirror images, 15; in knots, 16; in trefoil knot, 23; in chains, 48; deltalinks, 50; in polylinks based on snub solids, 60
Extensions of faces to support rings, 58, 63, 74

Figure eight knot: *see* Knots

Gordian Knot: *see* Knots
Gordius of Phrygia, 21
Granny Knot: *see* Knots

Hamilton circuits, 49
Hexagonal rings, polylink made of, 58
Human body, symmetry of, 15
Hunter, Dr. Edward, 17
Hunter's Bend: *see* Knots

Icosalinks, regular, 74
Interchange: for two highways, 2-4; for three highways, 3-7
Inversion symmetry: *see* Symmetry

Jayne, Carolyn Furness, 36

Kepler, Johannes, 68
Knots: figure-eight, 16, 17, 23; overhand, 16, 21; square, 16, 26; bends, 17, 18; Hunter's Bend, 17, 18, 28; classifying, 21; Gordian, 21; trefoil, 21; Listing's knot, 23; amphichiral, 24; cinquefoil knot, 24; granny, 26; bowline, 28; stevedore's knot, 28; Turk's Head, 28

Listing, Johann B., 23*n*

Matthew, Morton, 50*n*
Meshes in weaving, 11, 12

Nolink, 50
Nonperformable operation, 15
Nonredundant cloverleaf, 3

Octalinks, regular, 73
Overhand knot: *see* Knots

Pentagonal polylinks: 63
Performable operation: 15
Planes of symmetry: *see* Symmetry

Redundancy: 2
Regular polylinks: definition, 52; constructing, 53, 56
Rhombicuboctahedron: basis for polylink, 59
Rhombic solids: bases for square rings, 56
Rigidity of polylinks: 49
Ring of rings: 45

Scaling polylinks, 64
Sculptural diversity: *see* trefoil knot, 23
Snub polyhedra, as bases for polylinks, 60
Spherical approximations, 60
Square knot: *see* Knots
Square rings, polylinks of, 53
Stability of deltalinks, 49
Star hexagonal rings, polylink made of, 59
Star pentagonal polylink, 68
Stars of David, 59
Stellations: *see* extensions of faces
Stevedore's knot: *see* Knots
Sub-highways, 1
Symmetry, geometric: axes of rotational, 13; planes of reflection, 14;
    center of inversion, 16

Tabby, 9; interpenetrating, 10
Tait, Peter Guthrie, 24*n*
Tetralinks, regular, 73
Traffic circle, 4
Trefoil knot: *see* Knots
Turk's head: *see* Knots
Twill: 9

U-turn: 1, 2

Warp: 9
Weft: 9